From [obscured by barcode] the
Ant[obscured] [obscured]usy
days [obscured] [obscured]hows
reader [obscured] [obscured]me.
Each chapter covers a single bird during a
single hour, highlighting twenty-four bird
species from around the globe, from the tropics
to the polar regions. At night, we encounter owls
and nightjars hunting and kiwis and petrels finding
their way in the dark. As the sun rises, we witness the
beautiful songs of the "dawn chorus." At eleven o'clock in
the morning, we float alongside a common pochard resting
with one eye open to avoid predators. At eight o'clock
that evening, we spot a hawk swallowing bats whole,
gorging on up to fifteen in rapid succession before
retreating into the darkness.

For each chapter, award-winning artist Tony Angell
has depicted these scenes with his signature pen and ink
illustrations, which grow increasingly light and then dark
as our bird day passes. Working closely together to narrate
and illustrate these unique moments in time, Hauber and
Angell have created an engaging read that is a perfect
way to spend an hour or two—and a true gift for readers,
amateur scientists, and birdwatchers.

Praise for **BIRD DAY**

"*Bird Day* is a brisk, high-concept read. It lends the reader a pair of giant wings to soar across the globe, peeking each hour at the lives of the world's most fascinating birds. Author Mark Hauber is a research ornithologist, and the text often draws upon his studies. We meet cooperatively breeding superb starlings; a secretary bird that stomps venomous snakes into submission; a duck keeping one eye open while sound asleep; a bat hawk that swallows its nocturnal prey whole. Lushly patterned chiaroscuro drawings by Tony Angell heighten the mystery and delight of these tall-but-true bird tales."

JULIE ZICKEFOOSE, author and illustrator of *Letters from Eden*, *The Bluebird Effect*, *Baby Birds*, and *Saving Jemima*

"A wonderful book that simultaneously made me nostalgic about a cave full of oilbirds in Trinidad and a kiwi running between my legs in a New Zealand sleet storm—and further informed me about the lives of birds. A brilliant collaboration between a first-rate behaviorist and my favorite bird artist."

PAUL R. EHRLICH, author of *Life: A Journey through Science and Politics* and *The Birder's Handbook*

"As much a meditation as a book, Hauber and Angell's *Bird Day* gives us a bird to think about at each hour of the day and night. They take us around the world, visiting birds including the brown-headed cowbird (5 AM), Hauber's own research subject; the ocellated antbird (noon); the Cook's petrel in New Zealand (10 PM); and twenty-one others. The narrative brilliantly captures the moment; the art makes the moment come alive. *Bird Day* is an excellent pairing of text and art, one I will return to again and again as the hours go by."

JOAN E. STRASSMANN, author of *Slow Birding*

"One could not ask for two better field companions than Mark Hauber and Tony Angell as they observe some of the world's most interesting birds. Their book gives us all new ways of seeing, hearing, and thinking about them— hour by hour—without ever leaving home."

ROBERT MCCRACKEN PECK, author of *A Celebration of Birds* and *The Natural History of Edward Lear*

Bird Day

BIRD
DAY

A STORY OF 24 HOURS AND 24 AVIAN LIVES

WRITTEN BY	ILLUSTRATED BY
Mark E. Hauber	Tony Angell

The University of Chicago Press
Chicago and London

The University of Chicago Press, Chicago 60637
The University of Chicago Press, Ltd., London
© 2023 Mark E. Hauber
Illustrations © 2023 by Tony Angell
Published 2023
Printed in China

32 31 30 29 28 27 26 25 24 23   1 2 3 4 5

ISBN-13: 978-0-226-81940-2 (cloth)
ISBN-13: 978-0-226-81941-9 (e-book)
DOI: https://doi.org/10.7208/chicago/9780226819419.001.0001

Library of Congress Cataloging-in-Publication Data

Names: Hauber, Mark E., 1972– author. | Angell, Tony, illustrator.
Title: Bird day : a story of 24 hours and 24 avian lives / written by Mark
    E. Hauber ; illustrated by Tony Angell.
Description: Chicago : The University of Chicago Press, 2023. | Includes
    bibliographical references and index.
Identifiers: LCCN 2023013411 | ISBN 9780226819402 (cloth) |
    ISBN 9780226819419 (ebook)
Subjects: LCSH: Birds—Behavior.
Classification: LCC QL698.3 .H383 2023 | DDC 598.15—dc23/eng/20230327
LC record available at https://lccn.loc.gov/2023013411

♾ This paper meets the requirements of ANSI/NISO z39.48-1992
(Permanence of Paper).

Contents

6AM (SUNRISE)

Silvereye

(AUSTRALASIA)

35

7AM

Bee Hummingbird

(CARIBBEAN)

41

8AM

American Robin

(NORTH AMERICA)

47

9AM

Eclectus Parrot

(AUSTRALASIA)

53

10AM

Indian Peafowl

(ASIA, INTRODUCED
WORLDWIDE)

59

11AM

Common Pochard

(EURASIA)

65

NOON

Ocellated Antbird

(CENTRAL AMERICA)

69

1PM

Secretary Bird

(AFRICA)

75

2PM
Emperor Penguin
(ANTARCTICA)
79

3PM
Superb Starling
(AFRICA)
85

4PM
Common Cuckoo
(EURASIA)
89

5PM
Indian Myna
(ASIA, INTRODUCED
WORLDWIDE)
95

6PM (SUNSET)
Standard-Winged
Nightjar
(AFRICA)
101

7PM
Great Snipe
(EURASIA)
105

8PM
Bat Hawk
(AFRICA AND ASIA)
111

9PM
Black-Crowned
Night Heron
(WORLDWIDE)
115

Preface

What do birds *do* all day long? They need to
find a rich feeding ground, protection from
intruders and competitors, warmth for their
vulnerable young, and safety from predators. To accomplish these tasks, birds keep
busy most waking hours. Some even keep
one eye open while sleeping!

   This little book is the result of my
decades-long interest in avian behavior. You
will see that each chapter covers one hour
of the day and one bird, whether an Indian
peacock flaunting its flamboyant train to
attract mates or the common cuckoo fan-
ning its gray plumage to keep reed warblers
at bay as it lays a secret egg. By following a
different bird during each hour, we meet a
whole flock of interesting characters. In the

early morning, we watch nearly blind New Zealand kiwis hunt by smell for earthworm prey. An hour later, we are in the company of South American oilbirds, navigating the night and their dark cave homes using bat-like echolocation. Just before sunrise, we track with the brown-headed cowbird as it invades another bird's nest to lay its egg. What happens to this egg interloper? Only a few hours later, we watch the American robin discover and jettison the cowbird's egg from its nest. In the light of the noon sun, dimmed and filtered by dense foliage before it reaches the tropical forest floor, we follow the ocellated antbird, the true victor in an army ant colony's raid, devouring the insects fleeing the battle for their lives. I cannot think of a better way to spend a day.

Our assembled cast is diverse and geo-graphically varied—offering a sense of our planet's biodiversity and fragility. We will travel to every continent in just twenty-four hours, a true bird day.

MARK E. HAUBER
Berlin, Germany, and Urbana, USA

Artist's Note

This book elevates our thinking and expands our appreciation and understanding of birds. We can now see birds as extraordinary coexisting species making their ways through life with impressive timing and precision.

Rereading this book, I was struck by how each bird's actions are exceptional. The fact that these actions are performed throughout the bird's life make them even more remarkable. A reader might imagine a bat hawk, nightjar, or antbird repeating these foraging ventures day after day or night after night. Vacations are not an option.

I used pen and ink to portray our subjects—putting the focus on the dramatic nature of what Mark describes in the narrative. Mark and I discussed the project several times before I launched my imagination into the species' activities and environment. It was a delight to prowl about a New Zealand island's understory with the kākāpō, imagine the secretary bird pouncing on and throttling a black mamba snake in a parched African savannah, and join the ocellated antbird as it followed a column of army ants in a Central American rain forest. Of course it was critical to depict each bird accurately in form and plumage, but I also wanted to represent them in a fashion that conveyed my admiration for their exceptional strategies for sustaining their lives, at a particular moment and in intriguing habitats.

This book's vivid narrative provides the reader with a body of new and intriguing information. It is my hope that these drawings, developed through our collaboration, will be visual compliments that bring *Bird Day* fully to life.

TONY ANGELL
Seattle, USA

Bird Day

Barn Owl
Tyto alba

(WORLDWIDE)

Hearing the scurry of a vole on the forest floor, we see a barn owl emerge from the darkness to catch its prey.

Unless you summer in the Far North or South, midnight represents deep darkness for plants and animals, including humans. Some species certainly embrace the night; they rely on scents, sounds, and even Earth's magnetic field to find their way. But as most birds depend on sight, you might expect a sleepy, dark start to our bird day. No! Many birds thrive at this time, including owls that have evolved to orient themselves and hunt for their prey in the dimmest of lights. There is perhaps no better example than the barn owl.

Barn owls live pretty much everywhere except Antarctica. But do not take that to mean they are uninteresting. Barn owls can hunt in total darkness. They hear the subtle noises made by voles, mice, rats, and other rodents and can locate them as they rush through the nighttime leaf litter. This requires the owls themselves to be quiet— and they are, flying so silently while hunting that they are able to pick up the softest of sounds from below. Barn owls look relatively large, but their bodies are no bigger or heavier than that of a dove. They are just covered in extra-soft feathers to minimize noise during flight. Large wings and a light body also allow this quieter, slower flight.

With these wings, the owls can even hover as they pinpoint their potential prey in the ground cover.

Hearing is one thing, but how do barn owls locate their prey in the dark? Unlike many other owls that have symmetrical ears, the orientation of feathers on the barn owl's face helps direct sound toward their ears, which are located at different heights. The differing heights help barn owls better perceive subtle changes in the direction and strength of the noises coming from a mouse or vole. This allows them to hear and even create three-dimensional mental maps— matching a sound with the distance to its source, above or below.

Thankfully, tonight is dry. Night rain is the enemy of the barn owl. It dampens their feathers, making flight less quiet, and adds background noise as it falls on the leaf litter, masking the sounds of small mammals scuttering on the ground.

Rain hinders the hunter, but stars, the moon, and even artificial lights near human settlements can help. They provide visual cues during many a midnight hour for the barn owl's flight. In turn, as these owls' eyes are nearly twice as sensitive as ours, they easily pick up the slightest of mouse motions, whether illuminated in the moonlight or in total darkness.

Little Spotted Kiwi

Apteryx owenii

(AOTEAROA—NEW ZEALAND)

Something smells good. For the little spotted kiwi, it must be an earthworm wriggling in the dark.

On the two-hundred-hectare island of Tiritiri Matangi, off the coast of the isthmus of the large North Island (or Te Ika-a-Māui) of New Zealand (or Aotearoa) and some twenty miles northeast of Auckland (or Tāmaki Makaurau), live three unique groups of flightless birds. Two of them, our kiwi and

the little blue penguin (or kororā), only come out at night. This helps them avoid predators, as they can employ the stealth that darkness affords. (The third, the takahe, uses its blue-green plumage as camouflage to hide in grass and tussock fields.) Little spotted kiwis are from an ancient lineage of birds related to ostriches. The kiwis' giant cousins the moas—some species nearly twice the size of ostriches—no longer roam these islands because humans hunted the birds to extinction only one hundred years after their first arrival to these lands around one thousand years ago.

Little spotted kiwis themselves may seem like unlikely prey, but they too are threatened. They are too small to withstand the many mammalian predators—stoats, ferrets, and feral cats and dogs—that humans have introduced to New Zealand. Thankfully, the island of Tiritiri Matangi is a sanctuary, where conservationists have removed predators and keep all mammals away—except human visitors. In this newfound safety, kiwis can live longer than thirty years.

Like the barn owl we just met, little spotted kiwis spend most of the night hunting. The kiwis do not mind the dark because they can barely see even during the day. Instead, the nearly blind birds rely on nostrils at the tip of their beaks to pick up the scent of a tasty earthworm or weta, a type of flightless New Zealand cricket. Mostly, kiwis spend the night alone, snorting as they walk to clear their air passages of dirt.

But during the mating season, females and males get in touch! Perhaps it is best they cannot see well given their dusty and often louse-ridden plumage. No matter—the kiwis use their voices to attract mates. Sometimes they call in turns and sometimes they engage in duetting—one bird starts and the other joins to complete a series of repeated calls. This night on Tiritiri Matangi Island, we can hear both a kiwi duet and a solitary bird snorting as it walks across the dark forest floor.

The ultimate result of the mating song is the largest egg relative to body size among all bird species: a kiwi egg can weigh up to a quarter of the mother's weight. It takes a tremendous amount of the female's energy and time to build such a large egg. At most, she can lay one to two eggs per clutch. And, in turn, it is the male's duty to incubate the egg—he sits consistently on the egg in a ground burrow, day and night, for up to two months. The benefit of a large egg is a large chick. A young kiwi only requires a brief month of feeding before claiming independence, snorting along its own path through the forest understory.

Oilbird

Steatornis caripensis

(SOUTH AMERICA)

After emitting a series of loud clicks from the forest canopy above, an oilbird emerges looking for fruit.

From a quick glance, oilbirds can look like their cousins the nighthawks and poor-wills, but these birds are unusual. They are one of the only nocturnal birds that eats fruit exclusively. Living in the rain forests of South America, they particularly love the fatty fruits of palms, laurels, and wild avocados—the last of these are just small enough for the hungry birds to swallow whole. After digesting the delicious fruits, the oilbirds return the favor, dispersing fruit seeds in their droppings and helping new plant generations to grow.

To find their fruit, oilbirds rely on an exquisitely sensitive visual system. Their surprisingly small eyes have extraordinarily large pupils to collect as much night light as possible. In fact, their eyes are more like those of deep sea fish than other birds!

Although their food hangs from trees instead of scurrying beneath them, like the barn owl's rodent prey, oilbirds' feather soft-ness and long wing arrangements allow the birds to fly slowly and even hover in front of a particularly rich bunch of palm fruits. Fly-ing slowly also helps these birds to navigate in and out of long cave systems where they roost and nest.

Back to that clicking, which is even

louder now. Oilbirds are one of a handful of bird species that use echolocation to orient, even in total darkness. Instead of the ultrasonic clicks that bats and dolphins use to navigate their milieu, oilbirds use a human-audible series of clicks to find their way through the pitch-dark caves and locate their nest within the breeding colonies. The birds build their nests from feces—another use for their droppings—and place those nests above running water, providing additional protection from predators should they enter the cave system. This morning one potential foe (an eager group of humans) enters and is greeted with a torturous cacophony of cries and calls. No wonder on the island of Trinidad these birds are often given the name *diablotin* in French, as these "little devils" certainly emit a painful alarm screech.

To understand their name in English, we must consider the fat oilbird chicks. In fact, the chicks are so fat that they are heavier than their grown parents! When humans first encountered caves full of oilbird chicks, many boiled them down for their valuable fatty reserves to use for cooking and lighting. Thankfully, today several national parks in South America protect oilbird colonies.

3 AM
Kākāpō
Strigops habroptilus

Kākāpōs can fly! Or at least they can flutter. This one has just landed after climbing a bush's thick branches.

The kākāpō, the largest of the nocturnal parrots, can use its wings after all. Like the little spotted kiwi, the kākāpō is also endangered. Only about two hundred live today and only on a handful of small, mammal-free islands off the main archipelago of Aotearoa, the Māori name for New Zealand. Only conservation scientists and the occasional Māori titi- (or seabird-) chick collecting

parties are allowed to visit these islands.

Given the night that we have spent in the dark searching for these charismatic parrots, we might expect that the kākāpōs have also evolved to sense sounds and scents. And, indeed, male kākāpōs use long-distance calls, or "booms," to attract females from the valleys to their mountaintop booming sites. Calling and displaying in the presence of other males in a group is called "lekking," and males at the same lek compete for the females' attention and mate choice. After mating, the male kākāpōs prove deadbeat dads; they provide no help to the female to maintain the nesting cavity (hidden below a dense patch of rootlets or formed underground), to incubate the eggs, or to feed the chicks. These are all hazardous tasks since the islands where these birds live are often rainy and the soil and understory are wet and muddy. This makes it hard for the females to make their way on foot between fruiting rimu trees and the hungry chicks.

To find an abundance of the ripening fruits of the rimu tree—an event that occurs once every two to five years—female kākāpōs also use their keen olfaction. Experiments have shown that a hungry kākāpō can even sniff out fruits and nuts sealed in a box. Forget that old myth about birds not being able to smell because many (especially nocturnal species) can certainly pass a test of odor perception administered to them. But the kākāpōs have another special sense. Like all other parrots, hummingbirds, and many songbirds, kākāpōs have the ability to perceive ultraviolet (UV) light. But this does not seem of much use for night foragers. Perhaps in time kākāpōs will let us discover their so-far-hidden use for this wavelength of light.

Even for kākāpōs, there can be too much of a good thing. In years of low rimu fruiting, conservation scientists conducted an experiment. The scientists aimed to fatten up female kākāpōs with an all-you-can-eat buffet of nuts and pine cones. Why? They wanted to give them the body weight needed to carry chicks even when their natural food supply was sparse. And put on weight the females did! They also began to lay eggs in these years of artificial plenty. But something strange happened—these eggs hatched into mostly male kākāpōs. This was not what the conservation scientists wanted! To increase kākāpō populations, they needed more females. In subsequent years, mothers-to-be were kept on a moderate diet, and the chick population returned to a more equal mix of male and female.

Common Nightingale

Luscinia megarhynchos

(EURASIA)

When we meet this melodious nightingale, he has just returned to Berlin, Germany, after a three-thousand-mile journey from sub-Saharan Africa, across both vast desert and the Mediterranean Sea.

What a trip for a small bird about the size of a house sparrow. When migrating nocturnally, this species uses both the stars above and Earth's magnetic field to orient itself north in the spring and south in the fall. But this bird is known for more than its long-distance travel. The nightingale is one of nature's most talented musicians—and the inspiration for many a musical piece penned by human composers, too.

Singing some two hundred different tunes or "phrases" on average, a single male nightingale uses its vocal prowess to both woo and warn. Yesterday, as nightfall arrived and darkness set in, the male began to sing a varied tune to attract the females' attention. If one had decided to become his mate, he would have continued wooing her with his songs to assure siring the eggs to come. Nightingales have a reputation for keeping their pair-bonds exclusive as they coordinate parental duties. The female builds the couple's ground nest and incubates the eggs, while the male defends the nest and the territory from intruders and predators.

If you listen closely, you can sometimes hear the nightingale male singing and the female joining in to sound the alarm. Perhaps they have spotted a squirrel hoping to make a meal out of their eggs.

But in truth the nightingale pair-bonds are not as faithful as they seem. Past genetic analyses of chick and parent DNA in a nightingale nesting neighborhood showed that a male with a larger song repertoire attracted females from those males that had inferior song routines! In fact, it is risky for males to settle too close to each other because the denser the breeding habitat is, the more likely that a female will stray from her mate. Still, it takes two to tango (or extra-pair couple), and it is not surprising that many males seek out these opportunities because nightingales live an average of just one to five years (though researchers have recorded one living to nearly eleven!). These short lives mean nightingales only have a handful of summers to mate and pass genes on to future generations.

So, after spending the night wooing his female partner (and perhaps his neighbor's too), as sunrise nears, the male nightingale switches his tune and starts to advertise his position and prowess to the neighboring males to protect his territory. This serves to reduce trespassing by both the neighbors and wandering males who have not been able to secure a territory on their own. What sounds to our human ears like a sweet song may mean stay away!

Brown-Headed Cowbird

Molothrus ater

(NORTH AMERICA)

Rise and shine! The cowbird's covert work starts early, just as the sunlight sets the stage in the forest canopy.

Female brown-headed cowbirds first face a commute for their most important daily chore—deception. Unlike 99 percent of all birds, cowbirds are "brood parasitic": they lay their eggs in the nests of other species. This means that a female cowbird breaks stereotypes. She never builds a nest, never

incubates her eggs, and never feeds her chicks. Instead, she must find a host species, a soon-to-be-fooled foster parent, willing to accept the foreign egg.

This morning, in a remaining primary forest fragment amid the corn and soy fields of the Midwestern United States, we watch a cowbird make her journey from a safe roosting site in a dense reedbed, where she spends the night surrounded by other cowbirds (including her mate or mates), blackbirds, and starlings looking for a suitable nest. This is an urgent journey. Overnight she has formed an egg that she is carrying inside her, ready to lay. A female cowbird can lay forty to seventy eggs in a single summer, and finding the resources to form each one takes a lot of foraging for seeds and insects in grassy patches. She also must seek out the calcium to form the eggshell. If we had been here yesterday afternoon, we might have seen the female cowbird pecking at snail shells or scattered bones, which are excellent sources of this mineral, on the forest floor.

Finding a home for her egg is both a physical feat and a social game. She must avoid being spotted by the rightful nest owners as she lays her egg. One common host, the chipping sparrow, will abandon even her own sparrow eggs after seeing a cowbird atop her nest. She would rather start a new clutch than raise a cowbird chick. This might seem like stubbornness bordering on heartlessness, but it may be wise, at least genetically speaking. An unrelated cowbird often begs more vigorously for parental provisions than its nestmates, outcompeting the mother's own offspring for food. Other birds can identify and remove a cowbird egg from the nest. And some may even recognize the hatched cowbird chick and feed it less than its pro-portional share of grub.

It is no wonder other birds will attack a female cowbird when they spot her at their nests. Red-winged blackbirds are strong and large enough to fight hard, peck forcefully, and even draw blood from the cowbird's head and back while she is still busy laying her not-so-clandestine egg. Yellow warblers utter cowbird-specific alarm calls and attract the neighborhood's other potential hosts, including the redwings, to mob and attack the cowbird. These calls may encourage warblers to check their own nests for cow-bird eggs that they may then bury in the soft lining at the bottom of the nest.

Thus, our cowbird friend is in a hurry. She must fly directly to her selected nest and sneak in before the rightful owner returns to lay her own eggs. As the cowbird's luck would have it, many host species do not spend the night on their nest before their own clutch is completed. These birds are not being neglectful parents. Visiting a nest repeatedly might attract the attention of predators hungry for the tasty eggs. In most cases, the fewer the visits, the safer the nest. Except from cowbirds.

How does the cowbird know where she's headed so early in the morning? She cannot spend her time searching for nests, typically well hidden in dense vegetation at this early hour of the day. Instead, she relies on reconnaissance performed during the previous days. When the sun is out, she goes nest-searching, and she must remember the particulars of where suitable nests are located and where she can lay an egg without detection.

We humans have what is called "episodic

memory," a clear ability to keep track of where, when, and what occurs at a given time. Female cowbirds routinely engage in such mental time travel too. To accomplish this cognitive feat, female cowbirds have evolved an enlarged hippocampus, the brain region that is responsible for spatial memory. Male cowbirds do not have particularly large hippocampi—perhaps because they do not need to search and remember where host nests are located. Species related to cowbirds that build their own nests and incubate their own eggs, such as common grackles and red-winged blackbirds, also do not have these enlarged brain centers.

It is now five minutes to sunrise, and the cowbird does not have much time. Female warblers and thrushes arrive just after the sunrise to lay their own eggs. If they recognize a cowbird's egg, they may toss it. Still, some hosts may worry about retaliation if they reject the cowbird's egg. They are right to be concerned. Some female cowbirds are known to engage in "mafia" behavior. If the host removes the cowbird egg from a nest, the cowbird may return and destroy the remaining clutch of eggs, forcing the host to start and build another nest. This intimidation seems to work. The female cowbird often locates the new nest. And, having learned her hard lesson, the host is less likely to remove any new cowbird egg

Once she successfully lays her egg, the cowbird remains in the forest patch. After we leave her, she will spend the rest of her morning searching for other suitable nests and memorizing a path for tomorrow's pre-dawn journey.

6AM (SUNRISE)
Silvereye
Zosterops lateralis

(AUSTRALASIA)

Here comes the sun—and the song chorus it inspires in birds, small and large alike.

With the exception of Antarctica, songbirds inhabit all continents and terrains of the globe. Therefore, a walk in a forest patch around sunrise might yield few visible birds, but their songs can be heard everywhere. Birds perform a "dawn chorus" in nearly all seasons, but it is the loudest and most persistent in the months leading up to the spring breeding season in the temperate climates. The dawn chorus is one of nature's

gifts to the early-rising birdwatcher and professional ornithologist alike. It allows us to hear the full range of the vocal repertoires of different local species, even some typically quieter individuals. For instance, many songs uttered in the morning hours are never repeated during broad daylight.

Can you hear the silvereye? It inhabits coastal Australia and New Zealand's islands and belongs to a vast group of closely related species of wax-eyes or white-eyes (birds which all share a prominent, bright set of eye rings), which are found from Africa to South and East Asia and the Pacific Islands. Silvereyes came to New Zealand from Australia without human intervention. One theory is that severe storms blew the birds thousands of miles across the Tasman Sea that separates these landmasses. Until the arrival of human inhabitants around one thousand years ago,

New Zealand was mostly free of mammals (except for bats), and the endemic birds of New Zealand had evolved for millions of years in the absence of these nest predators. Not the silvereye! Its origin in Australia prepared this species to live on the main islands of New Zealand, which now are fully overrun by rats, mice, possums, stoats, and other invasive mammalian predators of both native small songbirds and larger non-songbirds. Therefore, the silvereye has been able to persist in most forests in New Zealand, and its song, including during the dawn chorus, can be heard throughout the archipelago.

But why do male silvereyes sing so loudly and so persistently in the mornings? To whom do they call? While the cowbird spends its morning on a deceptive quest, the birds who join the dawn chorus are all about honesty. By performing a song—which involves heavy physical activity in muscle contractions, persistent perching, and early morning cold tolerance—male silvereyes advertise their ability to forage successfully and build up energy reserves during the day prior to the dawn chorus. When scientists help the birds, giving extra mealworm food to fatten up select silvereyes, the huskier birds not only sing more songs but also more complex tunes on the following mornings. In turn, birds that do not have access to extra meals continue to sing simpler and shorter songs in the dawn chorus.

Singing is all well and good, but what happens next? To understand this, we must travel to an island on the Great Barrier Reef in Australia. Here, on the small landmass of Heron Island, and without mammalian predators, the same species of silvereye breeds successfully and at high densities. Pairs nest within visible distances from each other. If you are thinking like the nightingale we met earlier, you might see an opportunity for the birds to mate with neighbors.

As it turns out, young silvereyes on the island couple up early and remain in a seemingly strong bond for the rest of their lives. Genetic tests confirm this faithfulness. Specifically, after performing paternity tests for chicks and parents, researchers found no evidence of a single mismatch: all offspring genetically belonged with their apparent parents. Perhaps there is a reason for the lack of mate switching on Heron Island. Here inbreeding with cousins of cousins for many generations has led to little genetic diversity, so cheating offers little or no genetic advantage. As a result, pairs appear to be better off being faithful and cooperative as true partners.

Bee Hummingbird

Mellisuga helenae

(CARIBBEAN)

In the warming morning light of the Cuban forests, we see a bee hummingbird patrolling a blooming kaleidoscope of colorful forest flowers.

Holding the official title of the world's smallest bird, weighing about two grams total (about the same weight as a US dime coin), the bee hummingbird is unique even among its species kinfolk. Its size is what dictates its uniqueness, since size and

metabolism are related in many animals—at least in warm-blooded vertebrates, including us. The smaller the animal, the faster the metabolism. As such, the bee hummingbird, a visually guided forager, relies on its acute sense of flower colors, including those that reflect UV light to draw both bees and birds to a bloom's nectar stores. This bird must visit up to 1,500 such flowers to find enough food in the course of a day's foraging. Some hummingbird species, especially those living at high elevation sites where temperatures fall dramatically at sunset, save energy overnight by dropping their own temperatures to ambient levels and going into a nocturnal torpor. In this slumber, their heart rate and metabolism nearly come to a halt.

The bee hummingbirds' flower visitations might seem self-serving. They are after all looking for sugar-rich nectar to fuel their days. But the plants benefit too; their pollen hitches a ride on the birds, allowing the fertilization necessary for the plants to bear fruit and reproduce. Though some hummingbirds are specialists and have evolved beaks that can access particular plant species' flowers and nectar stores, bee hummingbirds are less choosy and visit a variety of forest flowers along their daily routes. As with all hummingbirds, this species has a protractible and long tongue that allows it to lap up energy-rich nectar fluids. If there should be a small insect or spider hiding in the flower, it too might become part of the meal, providing much needed protein for this hummingbird's diet.

And our friend needs all the sustenance it can get. Hummingbirds have one of the highest heart rates among all animals—over one thousand heartbeats, accompanied by over two hundred breaths, per minute. Imagine those statistics as part of a human's visit to a nurses' station! (The average human has rates of fewer than one hundred beats and twenty breaths per minute.) In turn, bee hummingbirds and their relatives also need a high number of wingbeats to stay aloft—about eighty beats per second to hover in front of a nectar-rich flower—but males can elevate these wingbeat rates to well over one hundred per minute when showing off for a female, buzzing and singing to her to catch her interest. In turn, as with all hummingbirds, the females take on all the responsibility for nest building, incubating a typical clutch of two eggs, and raising the chicks to independence.

American Robin

Turdus migratorius

(NORTH AMERICA)

The morning light will reveal an unwelcome cowbird egg in the female robin's nest, but she will not know about it for a few more hours.

American robins are late risers. While many songbirds complete their egg laying as soon as daylight breaks, robins take it easy. An early bird myself, as a professor of ornithology, I like to be in the field at daybreak. In the Midwestern United States, where I work, that field is often a tree farm with neatly planted rows of ornamentals offering a thinly hidden spot for arboreal robin nests. Here we watch and follow birds in the early morning. In turn, I have a student who is a night owl and prefers to conduct fieldwork in the afternoons. When I find a robin nest with a certain number of eggs in the morning, I mark each of them with a felt-tip pen and send this student the geographic coordinates to follow up later in the day. More often than not, the afternoon check shows a new robin egg in the same nest.

As a scientist, I am here for the cowbirds that we met earlier, and robins are one of more than 250 species that host cowbird eggs. Unlike over 90 percent of these hosts, robins are robust egg rejecters—grabbing and tossing out the cowbird's beige and speckled eggs. This is no surprise since the cowbird eggs are so remarkably different from the robin's own immaculate turquoise-blue eggs. Still, even with this distinctness, the robin hosts typically take anywhere between an hour to a day to find and reject the foreign egg. My early morning walks in central Illinois or upstate New York during my career have boiled down to an egg hunt to find cowbird eggs before robins do.

The question of how and when robins recognize cowbird eggs in the mornings has fascinated scientists for nearly one hundred years. One researcher in 1929 used fake cowbird eggs of different sizes and colors to assess how robins tell cowbird eggs apart. Another researcher repeated these experiments in the 1980s, and my own students did the same in the 2010s. Robins reject small, beige, and speckled fake eggs more often than large, blue, and immaculate model eggs. In other words, the more ways a fake cowbird egg differs from a robin egg, the greater the chance the robin will reject it.

It is not color alone that catches the robin's attention. When researchers added deep-blue painted twigs to the nests, all female robins meticulously removed these sticks from the nest. Fake eggs painted the same color remained in half of the nests. What is more, those females that tolerated the deep-blue egg once also tolerated it when exposed to it the second time, whereas those females that removed the first of these model eggs also removed the second one. It appears that some robins have a peculiar personality and are consistently more or less fussy than others.

Of course, some egg obsessiveness is not surprising—eggs are the cornerstone of birds' breeding attempts. Without eggs, there are no hatchlings, and without hatchlings, there are no fledglings, juveniles, or next generations. My lab's recent experiments showed that objects that are less egg-shaped, such as narrow cylinders or sharp-edged diamond shapes, are rejected at increasing rates by robins, even when painted the same robin blue hues as their own eggs.

Adult (and baby) cowbirds are smaller than robins, and anyone who has seen a brood of robin chicks knows that competition for parental feedings is extremely fierce and space is at a premium. In fact, 50 percent of cowbird chicks placed in robin nests do not make it to fledgling. Why then is their so much fussing about a foreign cowbird egg in the robin's nest? Raising genetically unrelated young is costly for the parents and for the nestmates. The robin is better off removing the cowbird egg early than waiting for the cowbird chick to perish among the larger robin brood. So, plunk.

Eclectus Parrot

Eclectus roratus

(AUSTRALASIA)

All dressed up, but nowhere to go, the colorful female eclectus parrot prefers to spend the day at home.

Bright red and blue, the female eclectus parrot offers a beautiful but rare sight. Rare because she barely ever leaves her nesting burrow. There is a good reason: large, sturdy, and dry holes in ancient trees are at a premium, even in the old stands of the tropical Australasian rain forests where this parrot species breeds. Once she moves into a nesting hole, the female eclectus parrot must fight to protect her home from other females—even from other species—and she spends most of her time at the entrance or at the bottom of the cavity. With a life span of up to thirty years (at least in captivity), a safe and sound nesting hole becomes a critical advantage for successful breeding.

What then is the use of her beautiful bright red-and-blue coloring? For one, when sitting at the entrance of her nesting cavity, these colors are surprisingly less conspicuous against the brown tree trunk for a predator visually searching for vulnerable prey. Compare her with the mostly neon-green colored males of her species. Their green serves as perfect camouflage in the foliage of the tropical rain forest. Look at one of these males now searching for fruits, such as those of the lolly wine and the watery rose apple, and fleshy, fibrous seeds, including the macaranga and cape tamarind. He will later provide these forages, partially digested for the female's convenience, by regurgitating them to share with the homebody female. She will in turn share some with the developing young, too.

Eating your mate's partially digested food might sound intimate, but this is not an exclusive relationship. Multiple males feed the female, and they often mate with her as well. It is common for two males to look after the female and her offspring at any given nesting cavity. This is perhaps another reason for the fierce competition for tree holes—a prime location gives a female not only shelter for her two eggs but also access to suitable males that can share paternal care duties. As such, the brighter coloration of the female may serve as a signal to other females of her superior fighting abilities when defending a nest cavity and to males of her ability to mate and reproduce. But how does an eclectus parrot see itself? As with the other parrots that we met earlier in the day, eclectus parrots' retinas allow them to see in wavelengths including UV colors. In these frequencies, male and female parrots look even more spectacular, but their beauty is hidden to hawks and other, mammalian predators, which cannot see in this type of light.

Let us also not forget the real estate itself—not all nest holes are created equal. Some are flimsy, with too-thin walls that may collapse in the heavy tropical rains. Others are sturdy and keep the bottom of the nest dry, even in the most intense of down-pours. When nest cavities are prone to flood-ing, female eclectus parrots are known to kill the younger (and often male) hatchling in their nest to improve the chances of the older (often female) nestmate. This infanticide might seem surprising: why lay two eggs but kill one? If food and access to safety are limited, raising a sole chick successfully can be a more fruitful strategy than watching both chicks perish in the rising waters of a nest hole.

10 AM

Indian Peafowl

Pavo cristatus

No wonder Darwin was obsessed. The morning means it is time for the peacock to descend from his slumber in the trees and meet a peahen, or two or more, who has already started her morning on the ground below.

In the morning's bright daylight, we can see the colors of these birds come to life. Males alert nearby females to their presence by calling and displaying their spectacular train. Leaving the safety of branches high in the trees' canopy where they spent the night, males descend to the ground, disperse, and begin their audiovisual displays. Although iconic, this bird remains full of mystery, both in its native Asian range and across new feral homes, including the United Kingdom, Germany, the United States, New Zealand, and elsewhere. Despite the long-standing interest from Darwin and other evolutionary biologists that followed, there are many questions that have yet to be answered about the species.

Contributing to the mystery is the pattern that few scientists have studied these birds in their original territories, their "natural ranges." What we know from watching them in their introduced habitats and sites is that brother peacocks band together, displaying their stunning trains in the vicinity of each other as if cooperating to attract interested females to their proximity—no matter who gets to mate with them. This may seem altruistic, but it offers advantages from an evolutionary perspective. Genetically speaking, it might be best to have children yourself. But if you are to remain childless, it is still better for your brother to have children than for a complete stranger to do so. Evolutionary biologists call this type of reproductive gain "indirect fitness," and it seems to explain why, for example, we as humans tend to give more and larger gifts to family members more closely related to us.

Another mystery is the peacock's puzzling train. First note that this is not the bird's tail but instead a collection of extra-long back feathers supported by sturdy, brown tail feathers. Sometimes a peacock's train can stand over six feet tall around the male's back. It is full of deep blue and green hues. Again, there is beauty beyond what our human eyes can see, including richness of color in the UV ranges. This is in part the result of tiny nanostructures present in the barbules of the of the so-called eyes of the peacock's train. Yet, here's the enigma: the peahen (much like humans) does not see in UV light! Nor do the peacocks or the main predators of peafowl. It seems a waste of a show!

Perhaps females are looking for the loudest and brightest male. He certainly is hard to ignore. Look at the way he makes his tail shimmer in the sunlight. Once she chooses a male, the peahen is incredibly faithful and downright possessive. Why? Peacocks do not play any part in parental care, from nesting through incubation to sheltering the chicks. They are simply sperm donors looking to sire the next generation of peafowl. But, for a peahen, selecting the most flamboyant male increases the chance that her sons will also have bright trains and become successful in mating themselves. Scientists call this the "sexy son" hypothesis. When a female returns the next year to the males' group (again called a lek) to mate, it is no surprise that she chooses the same male again and again. She carefully inspects him—for example, she often stands behind him to examine his cloacal area as if looking for signs of sexually transmitted diseases. Males with fewer parasites are more likely to be chosen as mating partners.

Why do females keep others away from their mates while peacocks are happy to play mating wingmen? Unlike the males, the females looking for mates are unrelated to each other. As such, it makes no evolutionary sense for the females to share the best male. Chasing other females away seems a successful (if spiteful) strategy for the peahen.

11 A M

Common Pochard

Aythya ferina

(EURASIA)

Are those pochards winking at us? Birds of a feather float together for both socialization and protection.

Ducks make for a tasty morsel, both at night and during day, whether you are a human hunter looking for a trophy and meal or a peregrine falcon swooping to snatch a slow-flying bird from a fleeing flock. In turn, ducks take to sleeping whenever they can, including during the brightest parts of daytime.

Because of the high predation rates, most ducks and other waterfowl spend much of their life in hiding and on alert. Females use camouflage colors to hide atop their nests, or they flock and float near each other on large bodies of water to minimize exposure to predators. There is also safety in numbers. Accordingly, ducks turn to flocking when they need to feed, move, or sleep. The more individuals that band together, the less likely it is that a predator will be able to snatch any single individual. Safety in numbers helps ducks and other birds to focus on what matters: competing with others of their own sex, attracting the opposite sex, and other quotidian activities, such as preening their plumage. When it comes to preening, bird beaks are some of nature's most effective combs, allowing their owners to keep their feathers clean of lice and mites, well-oiled to repel water, and neatly arranged to facilitate flight.

As we see, a duck can also sleep during the day. The most dangerous part of sleep is the inability to sense and alter one's behavior in response to potentially harmful components of the environment: predators, parasites, competitors, or even the shoreline. As a response, ducks have developed a special type of slumber called "unihemispheric sleep," in which one side of the brain is asleep while the other is awake and observing the environment. Half-brain sleep patterns are often accompanied by what scientists call "eye-peeks." Such peeks are quick, not more than ten seconds, allowing the other eye and half of the brain half to fully engage in resting mode.

What exactly do pochards do when attempting to sleep? We can see this clearly during a visit to the Danube River as it flows through the city of Belgrade, Serbia's capital. In the winter months, the Danube is home to many migratory duck species, including our pochards. Looking at the winking birds, we might expect that the bigger the flock, the fewer the peeks. Why bother peeking when someone else is on the lookout? It is actually the opposite! The more neighbors a duck has, the more time it spends peeking during each bout of observation. Perhaps the ducks are quite sensitive to bumping into neighbors. A crowded patch of water means more peeks.

Ocellated Antbird

Phaenostictus mcleannani

(CENTRAL AMERICA)

A trail of ants leads us to the ocellated antbird and, in turn, both the antbird and its avian companions to food retreating from the advancing army.

Despite their name, antbirds do not have an appetite for ants. Instead, these birds are ant followers; they can be reliably found near an army ant swarm off to attack or retreating from their daily foraging expedition. As the army ants move about,

diverse insect prey jump, fly, or otherwise try to escape certain death in the jaws of the approaching troops. That is when our antbirds—and other ant followers such as tanagers, antpittas, and woodcreepers—swoop in. By foraging at different heights and distances from the swarm, such multi-species flocks of ant-following birds are able to share this unique feeding opportunity and even defend their swarm and territory from rival multispecies flocks.

How exactly do you find such a troop of similarly motivated birds? Some learn the calls of antbirds or other birds that follow antbirds. This is the stuff of nursery rhymes: a woodcreeper following a wren, following an antbird, following an army of ants. Scientists call this strategy "heterospecific eavesdropping," where one species listens in on the vocal conversations of other species. This

eavesdropping is not limited to ant followers, as listening to others' conversations is
helpful for more than foraging. Some horn-
bills, for instance, eavesdrop for protection
by tuning in to the anti-eagle alarm calls of
monkeys, and yellow warblers eavesdrop to
avoid being duped, tuning in to the calls of
the cowbirds we met earlier.

Eavesdropping of course requires famil-
iarity between the listener and the caller.
This means that eavesdroppers need to speak
the "languages" of more than their own
species. One brilliant experiment used antbird songs recorded in one place (mainland
Panama) in a new locale, an island created
in the last century by the construction of
the Panama Canal where there are no more
antbirds. The eavesdropper species in the
new location did not seem to recognize the
calls—but their counterparts on the main-

land were attracted to the same antbird's songs. It is a bit like trying to eavesdrop on a conversation in French when you did not grow up hearing French. You might recognize that it is a different language, but you are sure to miss the meaning and, hence, the way to the best patisserie.

Secretary Bird

Sagittarius serpentarius

(AFRICA)

The heat of the midday sun brings out a stunning hunter and warms the blood of its slithering prcy.

In the tropical and subtropical African savannah, we are lucky to see the increasingly rare secretary bird standing tall above the flowing grassland. Look at its three-foot-tall frame, its long tarsus (or foot bone), lean body, long neck, elaborate face pattern, and crown. Its costume is reminiscent of a nineteenth-century European clerk's outfit, after which this bird's Latin genus (*Sagittarius*) is named.

If you consider the bird's name further, this time the species name *serpentarius*, you get another clue about its favorite meal. Cold-blooded snakes use the midday heat to raise their body temperatures and are most prominent but also quickest at this time of day. So how does the secretary bird catch these snakes at their top speed? Again, take a look at the bird's appearance and its long half-scaled lower leg—these birds stomp their prey to death. It is a dangerous game because they must avoid venomous snake-bites with each stomp. Their tarsus is covered in scales to protect them from snake-bites. Together with caracaras of the tropical Americas, secretary birds are just one of a handful of hawk relatives that pursue their prey on foot rather than on the wing.

Heat, however, is also an enemy of the secretary bird, especially along its westerly distribution near the Namibian deserts. With global climate change raising average temperatures and causing extreme highs, the typical secretary bird breeding season has become measurably hotter. This added heat interferes with the growth and development of the embryos inside their eggs.

Direct human interference has made the breeding habitat of the secretary bird even more vulnerable; with fewer undeveloped grasslands and increasing agricultural land use, people's growing presence disturbs this bird's peace atop its bulky tree nests. Human passersby often fluster and flush secretary bird parents from the nest, disrupting their ability to incubate or shade the priceless eggs. The sad result is unprotected nests and eggs baking and perishing in direct sunlight.

2 PM

Emperor Penguin

Aptenodytes forsteri

(ANTARCTICA)

At most latitudes, we would expect bright light in the early afternoon. But this is not the case in Antarctica as the winter—this penguin's breeding season—approaches.

As daylight's duration shortens, most north-temperate and Arctic birds begin their long migration toward the south to escape the cold winters. Emperor penguins, native to the Southern Hemisphere, also begin the first of their many marches as the austral winter approaches. Except, these penguins head in the opposite direction of the tropics and warmer temperatures—into the polar winter. It is on an inland Antarctic ice field where they aggregate, court, find a partner, and mate to fertilize their sole egg. Once the female lays her singleton clutch, she transfers the egg to the male, who takes it into his brood pouch, a feathered pocket atop the penguin's feet that retains the adult's body heat. He will incubate the egg here for the next two months while withstanding the elements of the most brutal Antarctic winter weather. Meanwhile, his female partner is on a hunting trip to feed herself and fill her crop—a storage pouch in the upper tract of the bird's digestion system—with krill and fish, future food for the hatchling. The chick typically hatches just days before the female returns, receiving its first meal in the form of the male's crop milk, a regurgitated slush rich in fat. Penguins are not alone in this behavior. Doves and flamingos also feed their young with crop milk.

At this time in the season, there is still no direct sunlight at any point of the day. You might blame the many Antarctic storms falling on the penguin colony, but the true culprit is the southern polar winter. After the females return, the males quickly take off, forming long rows of marching penguins. They are headed toward the nearest ice break and open ocean access. It is time for the male to eat as the female feeds and broods to transfer heat to their solo chick back at the colony.

Once the chicks have grown to about a third of the adults' size, they are ready to withstand the Antarctic wind and cold on their own. Sort of. The young form a team of juveniles, called a crèche, to keep each other warm while their parents are off fishing. When the parents return, the chicks call out loudly with unique voices. Parents recognize their chick and are willing to share the latest catch with their young. After a bit more time, the young are ready to "fledge." For most birds, ready to "fledge" means ready to fly, but for the penguin it simply means to live and feed on its own. If everything is on schedule, at this moment the ice sheet where the young have lived begins to break up, providing easy access to the polar ocean where they too can forage for aquatic food, such as fish, krill, and squid.

Like most of our bird day, the penguin's early life is perfectly orchestrated. The penguin chicks need enough time to grow and fledge before the ice sheet breaks each season. Humans, who have rapidly and dramatically changed the planet's climate, are now altering this long-evolved penguin schedule, causing ice sheets to melt early and hurting these chicks' chances of survival.

Superb Starling

Lamprotornis superbus

(AFRICA)

When you are superb, you are never alone. A flock of starlings form a colorful pack to protect a shared nest. Their feathers shine especially bright in the afternoon sun already starting to set.

Africa's superb starling earns its extraordinary name. From its bright orange belly and iridescent blue-green back and chest to its prominent white breast band and yellow eyes, superb starlings exhibit a rainbow of colors. They make for an even more vibrant picture in real life because you are unlikely to see these birds alone in their savannah habitats—they are gregarious in both the dry nonbreeding season (when insect food is scarce) and the wet breeding season (when food is plentiful). For this species, living in flocks also means breeding in groups. Multiple pairs lay eggs in a single bulky nest, often placed inside the safety of a thorny bush or tree. When the chicks hatch, the group works together to raise and protect the next generation of young.

Why should such cooperative breeding evolve in superb starlings in the first place? An increasing number of studies seem to show that this strategy might help protect starlings from unpredictable variability in weather patterns. Birds with more helpers fare much better than solitary pairs when an unexpected rainy season arrives and it is suddenly time to breed.

Helper numbers are also critical for increasing reproductive success in many other cooperatively breeding species, and, with superb starlings, females actively recruit unrelated helper males by allowing them to sire some of the young. What could be a better incentive for a male to contribute than the possibility of having his own offspring also in the nest?

Female superb starlings can seek new partners for other reasons. When their own mate and the available helpers are too closely related, females go outside the group to find new males. More diverse fathers means more genetic diversity in the progeny. This diversity in turn helps the birds' immune systems fight off pathogens and other diseases.

Given the important role of females in making decisions about reproductive success, it is not surprising that female superb starlings find themselves in conflict with other adult females. Though males might have many roles—as primary partner, helper, or a fling (in the name of genetic diversity!)—females only have one. Females who do not serve that primary role are often cast out of the group. Without a home, they are wanderers who may perish alone.

But what determines which females rule the roost and which must roam alone in the bright afternoon sun? Remember what makes the starling superb—that colorful plumage. Among the many starling species that live across Africa, the more colorful female starlings are more likely to engage in both sexual competition and cooperative breeding strategies.

4 PM

Common Cuckoo

Cuculus canorus

(EURASIA)

It is not just in the morning darkness that birds are duped. The cuckoo likes to deceive its hosts in the late afternoon sun.

By this point in the day, we have seen quite a few eggs. Why do all bird species lay eggs instead of live offspring? One explanation is that you cannot fly when you are pregnant. Bats, of course, would disagree. Another idea: evolution. When your own ancestors laid eggs instead of birthing live

youth, it is hard to evolve to do it another way. Yet snakes and lizards offer another dramatic counterexample, as "viviparity"—the birthing of live and mobile offspring—has evolved repeatedly among many of these originally egg-laying reptilian lineages, including sea snakes.

Whatever the reason, most birds lay eggs that are recently fertilized and undeveloped. One exception to this rule is the common cuckoo. The female cuckoo employs a similar strategy to the cowbird we met earlier, leaving her egg in another mother's nest. Like the cowbirds, cuckoos find a time when the host mothers are typically away from their nests and busy feeding themselves or defending their territory.

Unlike the cowbird, the cuckoo keeps her egg in her oviduct to be incubated for an extra day before she lays it in the nest of another species, replacing one of the host species' eggs. Why should cuckoo eggs be "preincubated" before laying? Preincubation gives the cuckoo chick a chance to hatch earlier. This early start spells doom for the host's clutch and brood, as the young cuckoo is a vicious killing machine. Within two to four days of hatching, while still naked and blind, the cuckoo chick pushes all the other eggs and chicks from the nest, often while helpless foster parents watch. These attempts to eliminate their competitors for parental provisions may go on for days. In one study, the cuckoo chicks kept tossing even as the researchers returned the evicted eggs to the nest again and again.

There are costs to this chick's attack. The more host eggs the cuckoo chick tosses and the harder it works to toss them, the slower the parasitic chick will grow. The time the chick invests in eliminating its competition might be better spent begging for nutritious meals from the foster parents. Still, the effort seems to be worth it. When cuckoos fail to eliminate host chicks and are forced to cohabitate, cuckoo nestlings grow slower and die more often than when living alone.

To avoid seeing another bird's chick rule their nest, some hosts, such as great reed warblers, act early to stop the female cuckoos from laying their eggs. These hosts are strong enough to draw blood from and even drown the female cuckoo while attacking it. After spotting a cuckoo, other hosts, including the Eurasian reed warbler, will become more sensitive to the possibility of a parasitic egg in the nest and discriminate against foreign eggs in the clutch, thereby annihilating the cuckoo's chances.

This enemy egg detection is harder than

it sounds. Cowbirds lay eggs that are proportional to the female parasite's size. Not so for the cuckoos, which lay especially small eggs relative to their own body size. This is part of the deception—the cuckoo's egg is similar to the foster parents' in size and appearance, if not contents!

Indian Myna

Acridotheres tristis

(ASIA, INTRODUCED WORLDWIDE)

Although the humans intended the trays of food pellets to feed feral cats in the evening sun, pairs of mynas descend on these trays for a feast of their own!

As humans change the face of the world, the Indian myna is happy to adapt. We have helped these feathered pioneers to spread from their native range in India and nearby regions and take over diverse new lands and habitats, from the coasts of the Pacific Islands to the Mediterranean Sea.

Who can forget the superb starlings of

Africa we met earlier? Indian mynas are the second species in the starling family that we have encountered today. In their invasive ranges—in Israel, for example—they are often seen stealing cat food left outside folk's doors for feral felines in the late afternoon. But we easily could have spotted them earlier in the day when they are loudly communicating with each other as pairs or aggressively defending their valuable nesting cavities from other species.

Mynas are some of the bravest and most adventurous birds in the world. This is made clear by their ability to invade several new continents and islands far, far away from their native South Asia. People have deliberately moved mynas around the globe because these birds make colorful, active, and friendly pets. But pet owners may have also lost or deliberately released their caged pets and given rise to new local populations.

In addition to living with humans as pets, mynas also perform work for some of us. Their appetite for insects makes them useful pest control agents in places like plantations easily overrun by arthropods. But again, some of these "working" birds have fled the proximity of plantations to establish new populations in more natural habitats where they compete with established, native birds.

For example, there are today few places on the subtropical half of New Zealand's north island or in coastal eastern Australia where mynas are not an everyday sight, fighting with people and other birds to protect their young chicks. Mynas harass native species and occupy valuable and rare nesting cavities, especially in Australia, New Zealand, and the Pacific Islands where native cavity-nest excavators, such as woodpeckers, are absent.

What makes a pioneer? How are the first mynas to enter a territory suited for success? Pioneer mynas are smarter and less fearful of novelty. It is true. When scientists compared mynas in their invasive sites in Israel with those in their native range in India, invading mynas were better at solving food-oriented puzzle boxes.

Today, mynas are still actively spreading to and invading new territories. In the Mediterranean, we can see these birds catching a ride on a commercial ship making an evening departure from the port of Haifa. Here in Israel, a release of mynas at a zoo led to a population boom nourished by cat food. We are seeing in real time the same spread beyond the city limits of Miami, Florida, as the birds move north and into suburban habitats along the eastern coastline.

Standard-Winged Nightjar

Caprimulgus longipennis

(AFRICA)

The male nightjar delights onlooking females (and lucky birders alike) with its twilight mating flight—when it is not busy sweeping flying insects toward its open mouth.

Feeding and breeding along the savannah belt of sub-Saharan Africa, the standard-winged nightjar attracts flocks of human birdwatchers during its breeding season. "Standard-wing" might seem a misnomer since we see two long flight feathers, one trailing each of the wings.

During the course of evolution, the benefits of an elaborate mating flight have changed the shape of male standard-winged nightjars more than many other birds. It is not the colors of its plumage or its extravagantly melodious songs that set this species apart but those elongated feathers set symmetrically behind each wing. Longer than its own body length, these special feathers come with a long, barbless shaft and a dark patterned feathery patch toward the tip. What a nightmare it must be to carry these feathers while foraging for food! They seem to drag the bird's otherwise light and airy flight, making it a struggle to chase moths and other flying insects in the vanishing daylight.

There ought to be some advantage to

growing these feathers. A special type of Darwinian natural selection called sexual selection results in traits—including unusual feathers—that can impart such an advantage. We saw this before with the peacock's train, and we can see it in other more distantly related species of nightjars, too. The peacock's exaggerated feather suite attracts the peahens' interest and mating, allowing males to pass on their genes and beautiful trains to their "sexy sons."

So it is no surprise that when a female is nearby the nightjar's feathers take on a new role and position. The male nightjar engages special muscles to pull the two plumes up to form vertical flagstaffs. Watch these flags (or "standards") flow in the rippling air generated by the male's slow and fluttery flight. He displays his plumage at its most extreme while performing an aerodynamically difficult feat—flying at slow speeds close to the ground where females may be hiding.

The risk involved with this air show is high because it makes the showy male vulnerable to attack, so it must be important in attracting a mate. These feathers make the males more conspicuous during its otherwise well-camouflaged rest in daylight, too. Perhaps because of this, males do not help incubate the eggs during the day, only at night. Still, comparatively, these males appear to be better parents than the peacock and male kākāpō, who refuse to engage in any paternal duties at all!

Great Snipe
Gallinago media

(EURASIA)

Great snipes prefer to feed at dusk and dawn, their brown-mottled plumage helping them to disappear in the waning daylight.

Despite their bulky appearance, snipes are strong fliers able to maintain fast directional flight during long-distance migrations. In turn, when disturbed by a ground predator, or perhaps us watching, they zigzag on their way from the ground upward and then toward another patch of wetland habitat. Their eyes are high on their heads, similar to the closely related woodcocks,

giving them an extrawide visual field so they can spot predators—and us—from most directions, including from behind.

Aerial displays, beginning just after sunset and lasting into the night, are also key to signaling fitness, both to attract the attention of females and warn off other males from claiming the best territories. We do not only watch snipes fly—we also hear them. As they fly in circles and then dive shallowly toward the ground, we hear repeated buzzing "whoops." Birds produce songs and calls using their vocal organ, the syrinx, but this is not what we are hearing. Many avian species—including manakins, hummingbirds, and snipes—engage in nonvocal "sonation," generating loud noises that carry a communicative function. During courtship, both male and female snipes let the wind move through their tail feathers during their shallow dives. The resulting vibratory sound resembles an extended "baaaahh." No wonder some call the species "sky goats."

After their loud and persistent sexual displays, the female chooses a male and the two birds form a proper bond, remaining on the same territory to nest and raise a family. Females build a lined nest, scraping out a shallow cup on one of the drier parts of the territory to keep the eggs safe from flooding. Her four eggs are well camouflaged and perfectly shaped to fill the bottom of the nest, pointed into the ground, sharp end down. The other end of the territory is typically a wetland, with soft and muddy spots for the snipe to immerse its long beak in while it hunts for earthworms, insect larvae, and snails. About three weeks later, after the female hatches the eggs, each parent takes on two of the chicks, dividing the brood.

This reduces the chance that the total brood will be lost and divides equally between each parent the investment they must make in protecting and guiding the now mobile young. Talk about not putting all your eggs (or in this case, chicks) in the same basket!

Bat Hawk

Macheiramphus alcinus

(AFRICA AND ASIA)

As the evening sunlight continues to disappear and the total darkness of night approaches, bats come out en masse from the safety of their caves and other hiding places. In the waning light, some raptorial birds have evolved to exploit this abundance of potential prey.

The bat hawk, with his bright yellow eyes, waits for the flocks of bats to emerge and disperse into the night. The bat hawk is an especially well-adapted predator of the small insect-eating bats. See him waiting on a perch near the cave entrance? As the evening approaches, he will down bats just like swifts or nightjars swallow insects.

In addition to having larger than standard raptorial eyes, bat hawks also sport the widest and largest bill gape; they are able to open their mouths wider than any other hawk (relative to their skull size, of course). They are also extremely agile fliers, able to capture their prey following high-speed chases and swift turns. And those large gapes serve them well as they simply swallow bats whole in midflight, gorging on up to fifteen each evening in rapid succession before retreating into the complete darkness of the night.

We have to travel to equatorial and other tropical regions in Africa or Madagascar and Asia or New Guinea to see this twilight diner, and in each place the bat hawk feeds on its namesake mammals: no matter the location or adaptation, these hawks are bat hunting specialists, restricting their feeding behaviors to about an hour before and after the emergence of the swarms from the bat caves.

Even the breeding season of the bat hawk is timed and coordinated with their prey mammals. During incubation, the male of the pair must feed both the incubating female and himself. He does this by chasing slowly moving and poorly maneuvering pregnant female bats. In turn, the fledgling hawks will benefit from growing at the same time as the bat pups; as novice fliers, the juvenile bats make easier meals.

Black-Crowned Night Heron

Nycticorax nycticorax

(WORLDWIDE)

We know when to look for night herons from their name alone, but we do not always know where to find them.

Medium-sized waterbirds, night herons roost in the safety of dense vegetation during the day. At night they forage in shallow waters or perched on branches overhanging open water. The exception is breeding season, when the increased energy demands of parents and needy nestlings alike require around-the-clock feeding. Daytime can be an especially productive and effective time for these birds to fish. Of course, the downside of daytime foraging is more competition, especially from other fish-hunters, including larger heron species.

We can see herons on most continents of the world—whether along a reservoir on a crisp fall day in New York City's Central Park, during the balmy summers of the Little Balaton National Park in Hungary, or in the freezing cold snow showers of the Patagonian springtime. Night herons breed in what are called synchronous colonies; their rookeries occupy midsized trees and bushes, such as those nesting in a dense aggregation in Tel Aviv University's Zoological Garden. Having the nests built, refurbished, or taken over at the same time; laying eggs synchronously;

and raising the chicks alongside the neighbor's chicks protects against predators visiting the heronry by creating a proportionally smaller chance of being the target of those predators. The night heron's eggs are immaculate and blue in hue, generated by the same pigment that colors the blue eggs laid by American robins and European song thrush and the green eggs laid by extant Australian emus and some of the extinct moas of New Zealand. (The pigment is called biliverdin, a component of the oxygen-carrying hemoglobin in the bird's blood, which is synthesized by the female rather than sourced from her diet.)

Night herons, though occurring widely and in large numbers, are threatened as humans develop their wetland breeding grounds into housing and shopping centers. As we consider places to build, we ought to look at and assess what other life is already using the territory around us.

Specifically, in the herons' case, where do these birds spend the nonbreeding season? Are they in danger during these nonbreeding times? One of the first studies attempting to examine the migratory behavior of black-crowned night herons was conducted at the National Zoo of the Smithsonian Institution in Washington, DC. Here, in the early 1900s, a scientist attached small colorful metal strips onto the foot bones of herons breeding within the grounds of the zoo. Bird bands became a pioneering research method, and the researcher heard reports of the banded birds from citizen scientists in the neighboring districts.

In the twenty-first century, we now use satellites and even cell phone towers to track birds. These birds can carry tags that are about the same sizes and types as those you might use to monitor the location of your stolen car or computer. Using such methods, we know that night herons fly as far as from the East Coast of the United States to Central America, the Caribbean, and the southeastern United States. Importantly, birds from diverse breeding sites returned to the same wintering locations repeatedly. This is evidence of both the complex migratory patterns of herons and the dangers of disturbing migratory species' homes during the summer and the winter.

Cook's Petrel

Pterodroma cookii

(AOTEAROA—NEW ZEALAND)

After darkness settles on Little Barrier Island (or Hauturu), about fifty miles north of Auckland, New Zealand, the calls and songs of the day's landbirds—the whitehead (or pōpokotea), the rifleman (or titipounamu), and the kaka—all fade. In their place we hear the noisy crash landings of the island's most common seabird inhabitant, the Cook's petrel.

During the breeding season, we see petrels in the hundreds of thousands, with each couple staking out a long underground tunnel. At the end of the tunnel is a nesting chamber housing a single egg and eventually a densely feathered chick. Small petrels like Cook's petrel benefit from arriving at the breeding grounds in the dark. For one, predators, including raptors or even related but larger seabirds, do not see their clumsy arrivals—crashing into the dense canopy of leaves, falling on the forest floor, and wobbling on their short and weak feet—and therefore cannot make them into a small meal. Once these petrels fall through the tall kauri trees, they use their olfactory skills to locate their nest, mate, and offspring before quickly disappearing into the burrow that they have dug and maintained over the years and breeding seasons.

Little Barrier Island is a unique place where these birds breed. From a distance, and even up close, it looks and feels like a volcanic island fit for a Jurassic Park movie sequel, as it towers above the rough seas of the Hauraki Gulf (or Tikapa Moana). It is just one of two isles on the globe where Cook's petrel populations survive as a species; the other site, Codfish Island (or Whenua Hou), is south of New Zealand's South Island (or Te Waipounamu) and off the smaller Stewart Island (or Rakiura). On Codfish Island lives a much smaller population of this petrel, numbering just around five thousand pairs.

Going back to the north, Little Barrier Island has in recent memory become the safest haven for New Zealand's land- and seabirds alike. This is because humans worked to eliminate the invasive mammals they had introduced. In the late 1970s, feral cats were removed from the island using traps and poisons, and in the early 2000s, the Poly-

nesian rat (or kiore) was eradicated using
GPS-guided helicopters that dropped poison
across the island's length and width (while
keeping native resident kiwis and tuataras
in holding pens until the poison dissolved).
As a result, the island hosts no more mammals, except of course the resident employ-
ees of the Department of Conservation and
permitted research crews. The island is also
home to giant wetas (flightless cricket relatives), kiwis (relatives of those that we met
this morning), tuataras (part of an ancient
reptilian lineage predating the evolution of
snakes and lizards), and other flighted and
flightless endemic inhabitants of the New
Zealand archipelago—including the New
Zealand storm petrel, which was rediscov-
ered in 2013 and has its only known nesting
ground on the island.

Cook's petrels occur along both coastlines of the Pacific Ocean, spending their summers and the breeding season in New Zealand on the west, then rapidly moving east toward the continental shelf of the Americas as winter comes to the Southern Hemisphere. There they feed on fish and squid, picking up prey from the ocean's surface or diving several meters into the water column in search of food. When the time comes, the Little Barrier birds complete a counterclockwise migratory return to New Zealand via the Hawaiian Islands and the equatorial Pacific. Genetic analysis of bird skins preserved in museum collections tells us that this movement pattern has been their migratory journey for a century or likely far longer. But with fish populations declining along the eastern coastline of the Pacific and invasive mammals severely hurting the petrel's ability to breed until recently, it is no wonder that Cook's petrels, persisting on just two breeding sites, are still considered vulnerable.

As with the other birds we have met today, Cook's petrel breeding is both complicated and risky. Parent pairs take turns to incubate the eggs. During any given bout, one of the parents must incubate the egg for several days while its mate heads out to feed, restoring the fat reserves required for another multiday incubation fast. Once their single chick is hatched, the parents increasingly leave the nestling alone while they hunt for prey to turn into a regurgitated oily-milky substance that will feed and fatten up their chick. They do such a great job that at the peak of their weight, the still flightless chicks can actually weigh 50 percent more than their parents! At this point the parents leave the chick alone. As the chick starts to drop the extra weight, hunger motivates it to become more adventurous. Eventually it will leave the nest burrow and, at last, take off from the island and begin adult life on its own.

11 PM

European Robin

Erithacus rubecula

(EURASIA)

In the dark of the winter night, we hear a tune that reminds us of earlier daytime hours and warns of warmer days ahead.

Nighttime, you might have thought, is reserved for owls, kiwis, nightjars, and other nocturnal birds. Yet, increasingly, birds whose colors and songs we associate with daytime brightness sing their song in the latter hours of the day. One of these birds is the European robin. This robin is typically an understory forest bird, hunting for insect prey in the dim light under trees and bushes.

Higher in the canopy, though, males like to find prominent posts and hold court with melodious songs for other robins to hear far and wide.

I ran into such a robin on a dark January night in northwestern Germany. The song was beautiful, and the little bird did not mind me filming his performance with my phone. But think about it—a robin singing in January in the cold of the continental winter! Why?

The robin is a migratory species in northern Europe, typically spending the cold season in western and southern Europe and North Africa. What, then, was this male doing with me close to the coast of the North Sea in winter? With snow on the ground and insect prey sparse, why was he here singing in darkness? Sure enough, I spotted feeders nearby with nourishing suet provided for insect-eating birds who were overwintering—food may not have been as sparse as I suspected. Still, I thought, he should have been curled up in the remainder of an old nest or other dense patch of vegetation, preserving precious energy for self-maintenance.

And this was January, still several months away from the onset of the breeding season. For whom was the robin singing? Surely it was not for a potential mate. Robins do not look for mates until springtime. Maybe his song was for neighboring birds, a warning to keep out of his wintry territory. He was claiming exclusive access to the resources of his own space. This was his food, shelter, and singing perch.

But why was the robin awake in the dark? He and I were in a Bremen suburb. Perhaps during the daytime hours the city's soundscape is simply too loud for the robin to compete with cars' background noise. Some birds, such as the great tits of Holland, solve this problem by raising their songs' pitch above the city's cacophony. But others, like robins, might shift the time at which they sing.

Another possibility is that the blue wavelengths of the city's streetlights confuse robins. Blue light is especially important to cue daytime and daylength periods for animals and plants. The same wavelengths emanating from the streetlights guiding me on my walk home might misguide the robin's song and his timing.

I wish him—and all the birds we have met today—goodnight.

Epilogue

There are many different birds that we could have met during our bird day. Indeed, there are almost an infinite number of configurations of twenty-four birds to fill twenty-four hours. Many different owls, several more kiwis, and at least one other fully nocturnal kite could have appeared as nighttime hunters in this book. The potential lists of diurnal birds are even more rich and diverse. There are distinct sets of birds to be met during the spring versus fall migratory seasons, during the winter or the dry season, during the summer or wet season, and in the temperate or the tropical zones. We might even have chosen to visit some of this same cast of birds at different times of the day or the year. In fact, when I conceived of this book, I originally considered following *each* bird for twenty-four hours.

I was fortunate to be based at the Institute for Advanced Study, in Berlin's Grunewald garden district, in Germany, while starting to work on the version of the book you are now reading. Here I saw and heard nightingales and robins. Because Berlin is known as the urban bird capital of Europe, with citizen science projects documenting the song dialects of local birds and a large population of bird-specialist goshawks hunting many of these (avian) residents, I could have written (and may still!) an entire volume on the Berlin birds you might meet at the different hours of the day.

Nevertheless, such a diversity of characters is only possible because there are more than ten thousand bird species that share Earth with us. Shockingly, however, for many of these species much of their behavioral diversity remains undiscovered. We know so little about their daily activities as well as their breeding biology. Even more sadly, each year we lose one or more of these species, some before we even first meet them, as climate change, habitat loss, and other human activities and needs take them away from our own and future generations. Therefore, this book is an urgent call for us to do everything within our means to put a stop to these devastating patterns and trends.

We have spent today together. Let us work together for all birds'—and the future generations of their fans'—tomorrow.

Acknowledgments

During the preparation of this book, I was generously hosted first by the Wissenschaftskolleg zu Berlin and then by the Humboldt Foundation Prize at the University of Bielefeld, Germany. I was also supported by the Center for Advanced Study and the Harley Jones Van Cleave Professorship of the University of Illinois at Urbana-Champaign, USA, a grant from the US National Science Foundation, and a grant from the US-Israel Binational Science Foundation.

I am grateful for being able to collaborate on this book with Tony Angell, who penned all these birds into beautiful drawings. And Joseph Calamia of the University of Chicago Press was the most helpful editorial guide that an author could ever wish for. Many colleagues, including Joan Strassmann and Marlene Zuk, kindly shared with me their experiences working on their own books. Danielle Allen kindly gave comments on the text, and Theresa Wolner provided indexing.

Finally, I am thankful for all the birds I have met during my life.

Further Reading

Ornithology. By Frank B. Gill and Richard O. Prum. W. H. Freeman and Company. 2019.

The Bird Way: A New Look at How Birds Talk, Work, Play, Parent, and Think. By Jennifer Ackerman. Penguin Press. 2020.

The Book of Eggs. By Mark E. Hauber. University of Chicago Press. 2014.

The Evolution of Beauty: How Darwin's Forgotten Theory of Mate Choice Shapes the Animal World—and Us. By Richard O. Prum. Doubleday. 2017.

Understanding Bird Behavior: An Illustrated Guide to What Birds Do and Why. By Wenfei Tong. Princeton University Press. 2020.

RELEVANT PEER-REVIEWED PAPERS

MIDNIGHT: Pena JL, DeBello WM (2010) Auditory processing, plasticity, and learning in the barn owl. *ILAR Journal* 4: 338–52.

1AM: Corfield J, Gillman L, Parsons S (2008) Vocalizations of the North Island brown kiwi (*Apteryx mantelli*). *Auk* 125: 326–35.

2AM: Konishi M, Knudsen EJ (1979) The oilbird: hearing and echolocation. *Science* 204: 425–27.

3AM: Aidala Z, Huynen L, Brennan PLR, Musser J, Fidler A, Chong N, Machovsky Capuska GE, Anderson MG, Talaba A, Lambert D, Hauber ME (2012) Ultraviolet visual sensitivity in three avian lineages: paleognaths, parrots, and passerines. *Journal of Comparative Physiology A* 198: 495–510

Hagelin JC (2004) Observations on the olfactory ability of the kakapo *Strigops habroptilus*, the critically endangered parrot of New Zealand. *Ibis* 146: 161–64.

Robertson BC, Elliott GP, Eason DK, Clout MN, Gemmell NJ (2006) Sex allocation theory aids species conservation. *Biology Letters* 2: 229–31

4AM: Landgraf C, Wilhelm K, Wirth J, Weiss M, Klipper S (2017) Affairs happen—to whom? A study on extrapair paternity in common nightingales. *Current Zoology* 63: 421–31.

5AM: Kilner RM, Madden JR, Hauber ME (2004) Brood parasitic cowbird nestlings use host young to procure resources. *Science* 305: 877–79.

Lawson SL, Enos JK, Mendes NC, Gill SA, Hauber ME (2020)

Heterospecific eavesdropping on an anti-parasitic referential alarm call. *Communications Biology* 3: 143.

Sherry DF, Forbes MR, Khurgel M, Ivy GO (1993) Females have a larger hippocampus than males in the brood-parasitic brown-headed cowbird. *Proceedings of the National Academy of Sciences* 90: 7839–43.

6AM: Barnett CA, Briskie JV (2007) Energetic state and the performance of dawn chorus in silvereyes (*Zosterops lateralis*). *Behavioral Ecology and Sociobiology* 61: 579–87.

Robertson BC, Degnan SM, Kikkawa J, Moritz CC (2001) Genetic monogamy in the absence of paternity guards: the Capricorn silvereye, *Zosterops lateralis chlorocephalus*, on Heron Island. *Behavioral Ecology* 12: 666–73.

7AM: Hainsworth FR, Collins BG, Wolf LL (1977) The function of torpor in hummingbirds. *Physiological and Biochemical Zoology* 50: 215–20.

8AM: Luro AB, Hauber ME (2017) A test of the nest sanitation hypothesis for the evolution of foreign egg rejection in an avian brood parasite rejecter host species. *The Science of Nature* 104: 14.

Luro A, Igic B, Croston R, Lopez AV, Shawkey MD, Hauber ME (2018) Which egg features predict egg rejection responses in American robins? Replicating Rothstein's (1982) study. *Ecology & Evolution* 8: 1673–79.

9AM: Heinsohn R, Legge S, Endler JA (2005) Extreme reversed sexual dichromatism in a bird without sex role reversal. *Science* 309: 617–19.

10AM: Petrie M, Krupa A, Burke T (1999) Peacocks lek with relatives even in the absence of social and environmental cues. *Nature* 401: 155–57.

Loyau A, Gomez D, Moureau B, Thery M, Hart NS, Saint Jalme M, Bennett ATD, Sorci G (2007) Iridescent structurally based coloration of eyespots correlates with mating success in the peacock. *Behavioral Ecology* 18: 1123–31.

11AM: Novcic I, Krunic S, Stankovic D, Hauber ME (2020) Duration of 'peeks' in ducks: how much time do pochard *Aythya ferina* spend with eye open while in sleeping posture? *Bird Study* 67: 256–60.

NOON: Pollock HS, Martinez AE, Kelley JP, Touchton JM, Tarwater CE (2017) Heterospecific eavesdropping in ant-following birds of the Neotropics is a learned behaviour. *Proceedings of the Royal Society of London B* 284: 20171785.

1PM: Portugal SJ, Murn CP, Sparkes EL, Daley MA (2016) The fast and forceful kicking strike of the secretary bird. *Current Biology* 26: R58–R59.

2PM: Aubin T, Jouventin P, Hildebrand C (2000) Penguins

use the two-voice system to recognize each other. *Proceedings of the Royal Society of London B* 267: 1081–87.

3 P M : Little J, Rubenstein DR, Guindre-Parker S (2022) Plasticity in social behaviour varies with reproductive status in an avian cooperative breeder. *Proceedings of the Royal Society of London B* 289: 20220355.

Rubenstein DR (2007) Female extrapair mate choice in a cooperative breeder: trading sex for help and increasing offspring heterozygosity. *Proceedings of the Royal Society of London B* 274:1895–1903.

4 P M : Birkhead TR, Hemmings N, Spottiswoode CN, Mikulica O, Moskat C, Ban M, Schulze-Hagen K (2011) Internal incubation and early hatching in brood parasitic birds. *Proceedings of the Royal Society of London B* 278: 1019–24.

Grim T, Rutila J, Cassey P, Hauber ME (2009) The cost of virulence: an experimental study of egg eviction by brood parasitic chicks. *Behavioral Ecology* 20: 1138–46.

Pyron AE, Burbrink FT (2013) Early origin of viviparity and multiple reversions to oviparity in squamate reptiles. *Ecology Letters* 17: 13–21.

Sulc M, Stetkova G, Prochazka P, Pozgayova M, Sosnovcova K, Studecky J, Honza M (2020) Caught on camera: circumstantial evidence for fatal mobbing of an avian brood parasite by a host. *Journal of Vertebrate Biology* 69: 1–6.

5PM: Magory Cohen T, Kumar S, Nair M, Hauber ME, Dor R (2020) Innovation and decreased neophobia drive invasion success in a widespread avian invader. *Animal Behaviour* 163: 61–72.

Magory Cohen T, Hauber ME, Akriotis T, Crochet P-A, Karris G, Kirschel ANG, Khoury F, Menchetti M, Mori F, Per E, Reino L, Saavedra S, Santana J, Dor R (2022) Accelerated avian invasion into the Mediterranean region endangers biodiversity and mandates international collaboration. *Journal of Applied Ecology* 59: 1440–55.

6PM: Fry CH (1969) Structural and functional adaptation to display in the standard-winged nightjar *Macrodipteryx longipennis*. *Journal of Zoology* 157: 19–24.

7PM: Avery M, Sherwood G (1982) The lekking behavior of great snipe. *Ornis Scandinavica* 13: 72–78.

Bostwick KS (2006) Mechanisms of feather sonation in Aves: unanticipated levels of diversity. *Acta Zoologica Sinica* 52S: 68–71.

Klaassen R, Alerstam T, Carlsson P, Fox JW, Lindstrom A (2011) Great flights by great snipes: long and fast non-stop migration over benign habitats. *Biology Letters* 7: 833–35.

8 PM: Jones LR, Black HL, White CM (2011) Evidence for convergent evolution in gape morphology of the bat hawk (*Macheiramphus alcinus*) with swifts, swallows, and goatsuckers. *Biotropica* 44: 386–93.

9 PM: Igic B, Greenwood DR, Palmer DJ, Cassey P, Gill BJ, Grim T, Brennan PR, Bassett SM, Battley PF, Hauber ME (2010) Detecting pigments from the colourful eggshells of extinct birds. *Chemoecology* 20: 43–48.

Scarpignato AL, Stein KA, Cohen EB, Marra PP, Kearns LJ, Hallager S, Tonra CM (2021) Full annual cycle tracking of black-crowned night-herons suggests wintering areas do not explain differences in colony population trends. *Journal of Ornithology* 93: 143–55.

10 PM: Rayner MJ, Hauber ME, Imber MJ, Stamp RK, Clout MN (2007) Spatial heterogeneity of mesopredator release within an oceanic island system. *Proceedings of the National Academy of Sciences USA* 104: 20862–65.

Rayner MJ, Hauber ME, Steeves TE, Lawrence HA, Thompson DR, Sagar PM, Bury SJ, Landers TJ, Phillips RA, Ranjard L, Shaffer SA (2011) Contemporary and historical separation of transequatorial migration between genetically distinct seabird populations. *Nature Communications* 2: 232.

Rayner MJ, Gaskin CP, Stephenson BM, Fitzgerald NB,

Landers TJ, Robertson BC, Scofield PR, Ismar SMH, Imber MJ (2013) Brood patch and sex ratio observations indicate breeding provenance and timing in New Zealand storm petrel (*Fregetta maoriana*). *Marine Ornithology* 41: 107–11.

11 P M: Fuller RA, Warren PH, Gaston KJ (2007) Daytime noise predicts nocturnal singing in urban robins. *Biology Letters* 3: 368–70.

Alert B, Michalik A, Thiele N, Bottesch M, Mouritsen H (2015) Re-calibration of the magnetic compass in hand-raised European robins (*Erithacus rubecula*). *Scientific Reports* 5: 14323.

Index

daytime, 65–66, 116, 129, 131–32. *See also* diurnal birds

diablotin (little devils). *See* oilbird (South America)

diurnal birds, 133. *See also* daytime

ducks. *See* common pochard (Eurasia)

dusk, 105. *See also* evening

East Asia, 37

echolocation, x, 15

eclectus parrot (Australasia), 52, 53–57

Eclectus roratus. *See* eclectus parrot (Australasia)

emperor penguin (Antarctica), 78, 79–83

emus, 117

endangered species, 17. *See also* extinction

episodic memory, 32–33

Erithacus rubecula. *See* European robin (Eurasia)

Eurasia. *See* common cuckoo (Eurasia); common nightingale (Eurasia); common pochard (Eurasia); European robin (Eurasia); great snipe (Eurasia)

European robin (Eurasia), ii, 128, 129–32

evening, 95, 99, 111–12. *See also* afternoon; dusk; night; sunset

evolutionary biologists, 60–61

extinction, 117; and climate change, 135; and habitat loss, 135; and hunting, 9. *See also* endangered species

fledge, ready to, 82

flightless birds, 7, 9–10, 124, 126

Gallinago media. *See* great snipe (Eurasia)

global warming. *See* climate change

goshawks, 134

grackles, common, 33

Great Barrier Reef (Australia), 39

great snipe (Eurasia), 105–9, 106

great tits (Holland), 132

guácharo. *See* oilbird (South America)

habitat loss, and extinction, 135

Hauturu. *See* Little Barrier Island (Aotearoa/New Zealand)

hawks, 76. *See also* bat hawk